THIS NOTEBOOK BELONGS TO:

____

Algebra Identities

Product of two binomials

$(x + a)(x + b) = x^2 + (a + b)x + ab$

$$\begin{aligned}(x + a)(x + b) &= x^2 + bx + ax + ab \\ &= x^2 + (a + b)x + ab\end{aligned}$$

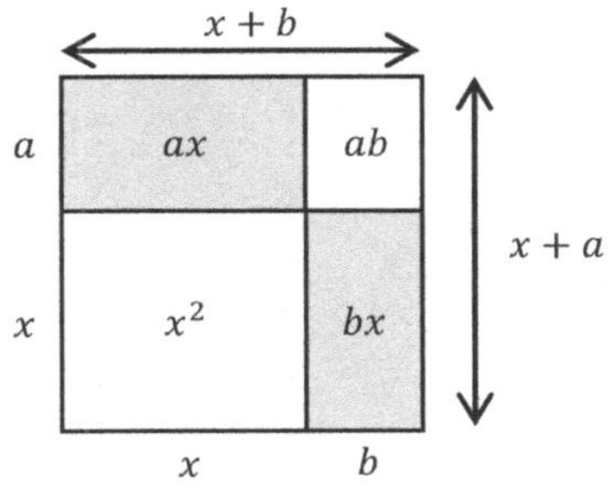

Product of the sum and difference of two binomials

$a^2 - b^2 = (a + b)(a - b)$

$$\begin{aligned}(a + b)(a - b) &= a^2 - ab + ab - b^2 \\ &= a^2 - b^2\end{aligned}$$

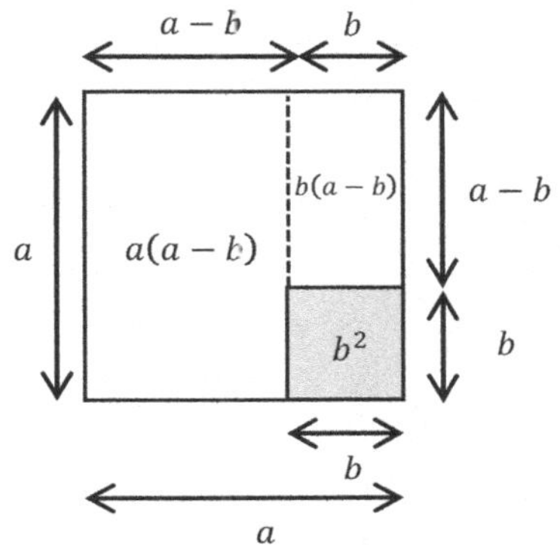

Square of the sum of two binomials

$(a + b)^2 = a^2 + 2ab + b^2$

$$\begin{aligned}(a + b)^2 &= (a + b)(a + b) \\ &= a^2 + ab + ba + b^2 \\ &= a^2 + 2ab + b^2\end{aligned}$$

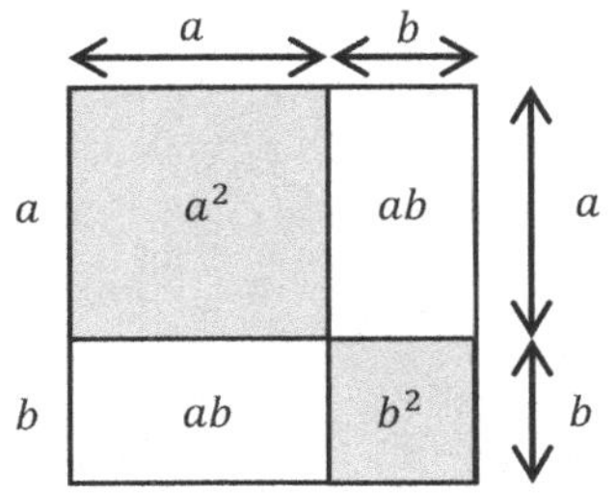

Square of the difference of two binomials

$(a - b)^2 = a^2 - 2ab + b^2$

$$\begin{aligned}(a - b)^2 &= (a - b)(a - b) \\ &= a^2 - ab - ba + b^2 \\ &= a^2 - 2ab + b^2\end{aligned}$$

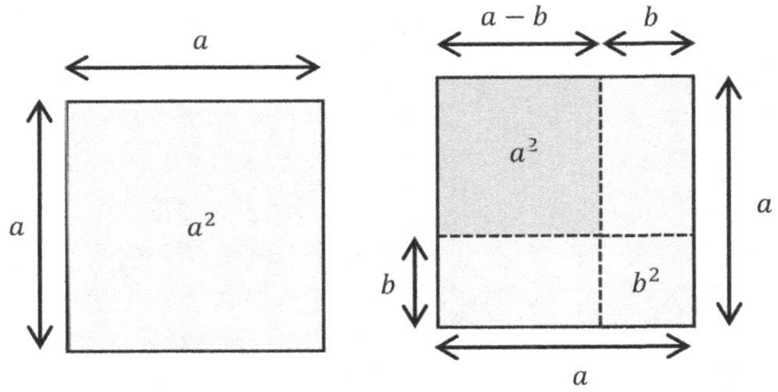

Difference of two cubes

$a^3 - b^3 = (a - b)(a^2 + ab + b^2)$

$$\begin{aligned}(a - b)(a^2 + ab + b^2) &= a^3 + a^2b + ab^2 - a^2b - ab^2 - b^3 \\ &= a^3 - b^3\end{aligned}$$

Sum of two cubes

$a^3 + b^3 = (a + b)(a^2 - ab + b^2)$

$$\begin{aligned}(a + b)(a^2 - ab + b^2) &= a^3 - a^2b + ab^2 + a^2b - ab^2 + b^3 \\ &= a^3 + b^3\end{aligned}$$

Made in the USA
Monee, IL
07 November 2026

67873791R20079